Dogs make great pets. They are lots of fun. But think hard before you buy any dog.

Be smart. Read about the many dog breeds. What is your dream dog? Only you can tell.

Some breeds love people and some are shy. Some need a long run each and every day. Some have thick coats that need brushing. And some dogs eat so much they cost a lot.

None are perfect when they are pups. Every puppy is a lot of trouble. Don't expect any puppy to be neat and clean. Please do expect to spend lots of time with a pup each day.

Maybe you are not ready for a puppy. That does not mean you cannot have a great dog. Year after year, people leave dogs at shelters. You can find a great dog there.

Please don't buy a dog at any old pet store. Some pet store pups come from places where they may not have been treated well.

Good breeders only sell a few pups each year. Their pups are very healthy and happy. Look! Here is your dream dog!

Train your dog very well. Teach him to sit and stay. Give him hugs and kisses every day. When you are done, you will have a best buddy for years to come.